THE OPEN UNIVERSITY
A Second Level Interdisciplinary Course
Art and Environment

unit 5
imaging and visual thinking

Prepared for the course team by Professor Kristina Hooper. College VIII,
University of California, Santa Cruz

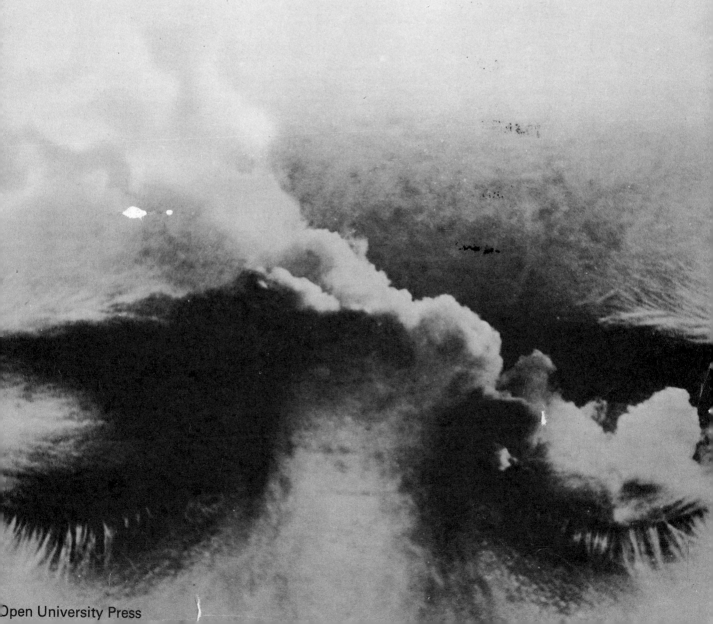

Open University Press

The Open University Press
Walton Hall, Milton Keynes

First published 1976

Designed by the Media Development Group of the Open University.

Produced in Great Britain by
Technical Filmsetters Europe Limited, 76 Great Bridgewater Street, Manchester M1 5JY

ISBN 0 335 06204 0

This text forms part of an Open University course. The complete list of units in the
course appears at the end of this text.

For general availability of supporting material referred to in this text please write to the
Director of Marketing, The Open University, PO Box 81, Milton Keynes, MK7 6AT.

Further information on Open University courses may be obtained from the Admissions
Office, The Open University, PO Box 48, Milton Keynes, MK7 6AB.

1.1

Plate 1   *Do you see the green discs as a group? Can you see them as a group? What about the other colours?*

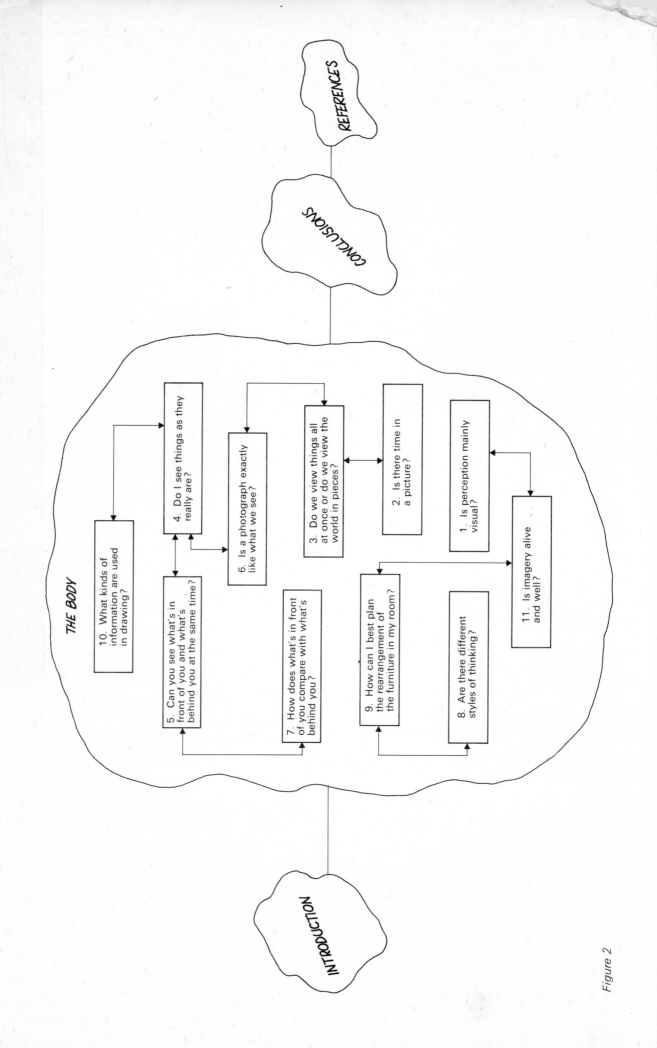

Figure 2

CONTENTS

INTRODUCTION

This unit is directed toward an understanding of selected aspects of thought processes. Specifically, it is directed toward an analysis of the content and functioning of the human mind insofar as these relate to the perceptual environment, mainly the visual environment.

The emphasis in this unit is on experiential projects designed for the organization of introspective insights. Consistent with this experiental approach, the orientation is active and constructive rather than purely meditative. Its attempt is to allow the reader to become aware of his or her own thought processes, such that he or she can (i) develop these processes, (ii) talk perceptively about thinking, (iii) play with thinking and seeing in different ways, viewing events from different perspectives, and (iv) create new objects and concepts based on new ways of viewing the world.*

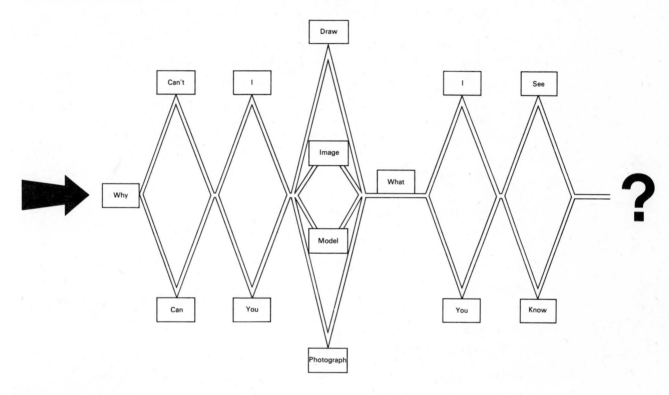

The goal of this unit is not to delineate current research on the topic of imagery or thought, but is instead to make the readers aware of the basic issues of this research at an experiential level so that they can pursue the topics of their own choosing at a later time. The goal of this unit is to arouse the reader's curiosity and to encourage a free flowing exploration of ideas. The implicit assertion is that with this sort of material, a highly structured and formal approach is inconsistent with these ends. Therefore, the basic approach of this unit is playful, rather than formal and rigorous (of course play has a structure also, but...). Students interested in the formal concepts and experimentation relevant to the topic areas covered in this unit are referred to the references listed throughout each section and at the end of the unit.

Figure 1   Begin at the left of the arrow. Move to the right, toward the '?', choosing any path

*In this introduction both the male and female pronouns will be used. In the sections which follow, 'she' will be used in the odd-numbered sections and 'he' will be used in the even-numbered sections.

Structure of the unit

The unit consists of an introduction, a body, a conclusion and a set of references. The body of the unit consists of eleven separate sections, shown in Figure 2, which, though numbered for identification purposes, are not organized in any predetermined order. These sections can be read (and thought about) (and worked with) in any order, depending on the interests of the individual reader. Unfortunately, though all the ideas in the sections are related, only one section can be read at a time. The reader must therefore make a decision at a basic level regarding how he or she will proceed. Sections which are closely related to one another have been connected in Figure 2 so as to assist the reader in the organization of his or her approach. However this should serve only as a guide to possible orders.

Each section has the same structure. This is shown in the following chart. It is initiated by a question which is stated in different ways. A general discussion section follows the questions, a discussion directed at an experiential exploration and logical organization of the issues involved. The set of projects which follow the discussions are the most important parts of each section. They allow the reader to become directly involved in the ideas being considered.

Question   What is this unit about? or
What should I expect to find in the following pages? or
What do I have to do?

Discussion   This unit is directed toward an understanding of selected aspects of thought processes ... alive and growing.

Projects   Read this unit.
Become actively involved in the ideas presented. Experience the exercises.
Become perceptive in your internal and external worlds.

An important aspect of each section is that it can be changed by the student. It can be expanded in many directions depending on each student's own thoughts and reactions. Each individual's experience will be different as a function of what he or she knows, what he or she expects, and how completely he or she processes the information available. Similarly, this entire unit can be expanded in the future as a function of reader input. There are now eleven sections, but there can be many more. Students are encouraged to raise their own questions, and to develop their own discussions and projects. Similarly questions can be raised which others (including this author) can discuss at another time. This unit should be alive and growing.

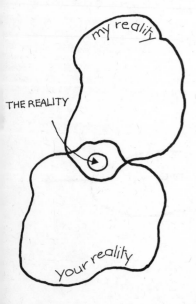

Preface

Thinking ... Visualizing ... playing with ideas ... spacing out ... spacing in ... looking closely at the world around you ... pretending you're standing on your head ... what do you see then?

Walking on the ceiling ... upside down like a fly ... Hmmm ... how is it now that a fly lands??
Does he land like a helicopter, or more like a Boeing 747 ... Hmmm ... the fly I see landing is a lady fly I think ... now how do I know that ... maybe it's the way she wiggles ... Can you see a fly wiggle??? In your mind I mean ... It doesn't matter to me if you can find a fly to watch wiggle ... and it doesn't even matter if flies even do wiggle ... I wonder if you can see a fly wiggle in a picture in your mind ...
CAN YOU?

Can you see the legs on the fly? Can you see its wings? ... Can you see its neck (Do flies have necks?) Does your fly (in your mind) have red tennis shoes on? If it doesn't, put red tennis shoes with pink stripes on your fly. (Have you ever before seen red tennis shoes with pink stripes? Does it matter??)

WHOOPS ... My fly in my mind just landed on my nose (in my mind) ... That really tickles ... I feel that I might sneeze ... Can you feel your fly tickling your nose??

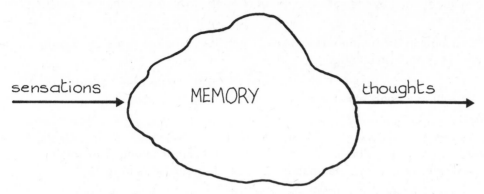

I wonder if our flies know each other ... Maybe they are old school friends ... Tell you what ... If you imagine your fly going to the centre of the room, I'll imagine my fly going to the centre of the room, and then they can meet and chat ... READY ... GO ...

Can you hear them talking?? I can barely make it out, but I think they're talking about meeting each other on the top of the Tower Bridge last Spring. Is that what you hear? Oh oh ... my fly is sweeping off ... and heading for disaster in a bowl of milk on the breakfast table ...

Whew ... close call ... this tall man turned the bowl over in his mind just before my fly landed ...

Ah ... settle back in your chair, or your sofa, or on the floor (Hmmm it's hard to imagine just where you are and how you're sitting, or lying, or standing ... Can you imagine what I'm doing as I write this ... do you have any idea of where I am ... or what I look like ...)

Now take a deep breath and centre yourself ... let the thoughts of the day flow through ... let them go into the infinite emptiness of your room ... there's plenty of room there for a whole lifetime of ideas ...

A room will not fill with ideas as a cup with tea ... (is there a cup in your room? I can see it there if I want ... can you?? I can even see it there without knowing whether the cup is blue or red or any other colour ... Funny a cup without colour ... hmm ... I don't even think it has any shape ... has ... my imagination sure does cheat on me).

AGAIN ... take a deep breath ... get the cup out of your mind ... whoosh ... a clear head ... imagine your head as an empty fishbowl ... now put some water in it and let the water move back and forth ... like the ocean ... soothing your forehead ... and the top of your head ... and the backs of your eyes ... close your eyes, and when you're centred, open them and begin reading again ...

Look at the world around you ... look at the objects, the angles, the colours, the forms ... look as though you're just awakened from a long long nap ... a nap which lasted many years ... a nap which has left you young and refreshed and alive ...

Really see the world ... in its complexity and its simpleness ... in the structure of the house's walls and the curve of the chair or the spoon on the table ... Take your time ... caress each form with your mind ... really feel it ... come to know it as a special friend, and then kindly bid adieu and move on ever so slowly ...

Now hear your world ... listen for your own breathing ... feel your lungs moving in and out ... feel your chest expanding ... control this ... make an art of it ... breathe with graciousness and style ... like a dancer strolling across a stage ...

Listen for the world nearby ... What do you hear? What don't you hear? Are you surrounded by the stillness of night? Or the joyous movement of a sunny day??

What's going on in your mind's ear? ... give it some more space by closing your eyes and listening ...

Hmmm ... taking time to know the world ... that can be fun ... taking time to play with our concepts of reality ... that can make one smile ...

What is there out there in the real world — in the physical world — that accounts for my view of it. Is it the molecules moving about, or the reflectance of the light, or the vibrations in the air? Is it the neurons firing in my brain, or the creativeness of my imagination? Is it all of these? None of these?

Is my view of the world the same as yours? I wonder. I hardly know my subjective reality, how can I know yours?

Let's meet halfway across the room, flying upside down, and let's chat in mid-air about our realities, like our fly friends.

Whoops ... I can't get halfway across the room ... my legs are stuck in some gooey fly paper ... Ah, I'm bound up in the limitations of the printed page ...

Tell you what ... I'll fly as far out into the room as I can and you try very hard to come the rest of the way to meet me ... I know that will be hard, but it's really important if we are to get a chance to walk around in each others minds ... if we are to share our concepts of reality ... if we are to discover how the human mind interprets the world ...

Let's learn to trust the other in our minds, and learn to move about each others' heads with great care. We must not joggle anything abruptly, yet we must leave the opportunity to change awarenesses and mental structures, for ourselves and for each other. We must learn to be open ... and honest ... and we must come to enjoy being silly and playful and oftentimes wrong ...

If our views of the world always agree, and if they always match physical realities, our discussions will come to be very, very, very boring. We might even come to realize that neither of us can really fly well upside down, and we'll come crashing to the floor ... (Hmmm I wonder what it would be like for a fly to crash to the floor? Would she go down straight, or would she float a bit?).

Hmmm ... imagine your fly again ... Does it have any legs at all?? Or any wings?? If it doesn't and you can say so ... or if you can accept the vagueness of your image to something to play with ... or if you can concentrate on just what your image looks like rather than what it's supposed to look like ... then you're ready to begin this unit ...

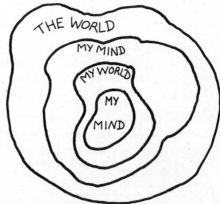

For unless you wonder about how your mind works, and how your friends' minds work, and unless you're interested in actively exploring your perceptual world, you'll become bogged down and lost in this unit ... You'll confuse simplicity with triviality, and you'll interpret your impreciseness in imaging or drawing or seeing in the terminology of failure rather than discovery.

Again, settle back, empty your mind of constraints, and begin to play with your view of the world. Question it. Turn it upside down. Move through the unit ... Move through your mind. Go forward. Go backwards. Begin at the beginning. Begin at the end. Think about imaging. Image about thinking. Imagine the time passing as you travel in your memory. Move backwards to your past. Move forward to your next perception. Put it all together. Take it apart. Read actively. Think actively. ENJOY!

I can't explore your mind for you ... you must do that I can only be your guide ... My ideas will exist in your reality only if you allow them and my past thoughts can merge with your present only if you leave me space ... space to know the questions and the uncertainties so that the answers (and non answers) make some sense.

KH

BODY

1 Is perception mainly visual?
or
How do the sensory systems interact?

Figure 3

We experience our environments as sets of objects which are located in certain places, which have basic attributes such as colour, shape, and size, and which have meaning in terms of past experiences and future expectations. In order to develop these perceptions, the nervous system must perform complex analyses of information in the environment, and it must combine this information in specific ways which are adaptive for individuals in responding to this environment.

The nervous system's access to basic environmental information — the data which will provide the basis for analysis and construction — are the sensory systems. Each sensory system (vision, audition, taste, smell, touch)* responds to specific kinds of physical energy in the environment, and these responses allow for interpretations of this environment. The nervous system does not respond to all types of environmental energy, and it does not respond to the entire range of energy available (Anderson, 1975).

*The proprioceptive system — which deals with perceptions of internal states, the location of parts of the body relative to one another, the feeling of movement of body parts, and the feeling of gravity — is not considered here.

10

For example, the visual system responds only to a very small set of wavelengths in the electromagnetic spectrum. It does not respond to X-rays or radar. Similarly, the range of auditory sensitivity is generally limited to between 20 and 20 000 Hz, and the receptivity to the different frequencies varies. There is no sensory system which responds to any range of magnetic energy.

Just as the human sensory systems do not respond directly to all the physical energy in the environment, integrative mechanisms at higher cortical levels do not treat responses from the basic sensory systems equally. Certain responses are attenuated and others are amplified. At the level of the visual cortex, edges of surfaces are amplified, as are lines at certain specific angles. At higher cognitive levels, other specific kinds of sensory information are selected in perception. In the interpretation of groups of objects, for example, proximity may be the main determinant of groupings rather than colour or other sensory attributes (though colour may be the basis of groupings in other cases). (See Plate 1.)

Similar reasoning applies at a cultural level, where we see that visual information is generally considered the main sensory data in an environment. In most industrialized cultures, taste, smell, and touch consciously play only a minor role in the assessment of an environment. Few distinctions are made, there are few deliberate sensory responses in non-extreme instances. Even one's auditory system plays a secondary role in modern societies.

Visual media provide the major input to our culture, and we therefore have a civilization which is mainly attuned to this modality (McLuhan, 1964), and one which regards mainly visual forms as art. Even dance, which is experience kinaesthetically by the performers, is experienced through the visual senses by our participant culture, and in this sense regarded as art.

An analysis of the English language provides additional evidence for the consideration of visual sensation as a primary mechanism of perception:

> '*See* what I mean.'
> 'Of course we can *look* at the problem in another way.'
> 'That's an important *insight*.'
> 'From this *perspective* it *appears* that you are correct.'

And, there is a large vocabulary within the language to describe many differences in visual attributes,* whereas there are few words to describe differences in the other senses.

An examination of the research on perception also supports the importance of visual processing. Mechanisms of the visual system generally serve as models for all sensory systems, as psychologists seem to study them more than the other sensory mechanisms, and they are generally the subjects discussed in the first chapter in a book on sensation and perception, and are the main topics of articles and journals dealing with perception.

The structuring of our environment is greatly affected by this general emphasis on visual information as architectural decisions are often based on mainly visual considerations. The architect should design the building so that it is structurally sound, so that it is *not* noisy, so that it is *not* smelly, so that the air does *not* taste peculiarly, and so that it is *not* too spread out or uneven.

However, visual criteria provide many of the constructive guidelines for a project. The architect attempts to create a building which is exciting, inspiring, classy, majestic, harmonic, beautiful, and pleasurable. Even though the functional aspects of buildings are of utmost importance, and the non-visual aspects are critical, the visual aspects are often those which are stressed. It is

*In fact our language affects our ability to discriminate differences in visual sensations. See Whorf (1956) and Lenneberg (1967).

the appearance of buildings which generally makes a building memorable, popular and accepted by most people. There are of course people who live in buildings, who pay attention to the functional aspects of these buildings. Yet these people respresent only a minority of all the individuals who will come in contact with a building.

Besides greatly affecting the physical construction of our environment, visual information seems to play a predominant role in the creation of impressions about this environment, as it is important in our assessment of non-visual as well as visual attributes of the environment. This is possible because of the correlation of events in the environment and memories for these correlations. Visual information, for example in a picture, can provide the observer with information about sounds, tastes, feelings and smells. If an observer sees trees blowing, for example, she can infer that it is windy. A well-done movie of a stable will evoke the smells of hay and horse manure. A vivid movie of a snowy day will cause the movie viewer to shiver.

However, though predictions can be made from only visual information, it is not clear that the interpretation of an area is the same if certain senses are not included in an analysis. Though the visual system is of primary importance in perception, the senses must be considered as perceptual systems (Gibson, 1966), which provide different sets of information which can be combined according to certain situations.

To see the importance of non-visual attributes in an environment, and the nature of the interaction of sensory systems, consider a deaf person's perception of an urban area. Imagine what a city would seem like without the sounds of trollies going by, without the chatter of street vendors, without the buzzing of crowds, and without the roar of cars. It may be pleasantly peaceful for a while, but then it may well become lonely and alienating. To know the pleasure of silence, one must experience non-silence. Though the visual sense does provide extensive information about an area, one must consider other perceptual inputs in the evaluation of an area.

An analysis of the perceptions of blind people also points to the importance of the inclusion of all the senses in the perception of environments. Think of a blind person standing on a high cliff above an ocean beach at dusk. She can feel the exhilaration of the place. She can experience the wind in her face, the sounds of the roaring waves, and the fresh air. However she is unable to experience many other aspects of this 'viewpoint'. Though she is able to experience those things which are close to her, she is unable to perceive very much distant information. She cannot see the people walking on the beach, or the ever-changing sunset, or the porpoises swimming parallel to the shore, or the multi-coloured sandstone on the cliffs.* However, the blind person on the cliff may have exactly the same general impression of this place as would a person who could see. She may find the place exciting, enjoyable and stimulating just as the sighted person. She may find the same position best for taking in the scenery, or for picnicking or for resting. Though the information used to develop these perceptions is very different, the general perceptions may be much the same. This may be because the sighted individual is unconsciously using a great deal of non-visual information in her judgments, even though she might assert that visual cues are not important in her perceptions. Again, it is the interaction of sensory information which is important in the interpretation of an environment.

* My thanks to Marigay Graña, with whom I spent many pleasurable hours on the cliffs near the beach in San Diego studying the perceptions of blind people. My thanks also to the blind individuals who accompanied us, who generously shared insights concerning their perceptual experiences.

PROJECTS

1   Take sixty seconds for each of the senses and list all the things in your environment that are perceived through this sense.

Vision   Hearing   Taste

Smell   Touch

What category has the most items listed? Which are the easiest to think of? Discuss the interaction of the different senses in terms of the specific items you have listed.

2   Find a picture of a complicated scene, for example a magazine picture of a group of people or a picture of a landscape from a calendar. Have three friends describe the scenes in terms of the visual, auditory, touch, smell and taste elements.

Example 1

It's a pretty place. the trees are especially fresh and alive. the dew on the leaves makes it seem like morning. Quiet, it's so very quiet in the forest in the morning. I wonder if there are any birds chirping, or maybe some small animals foraging around in the grass ... Smells. well it probably smells fresh like forests

KH 3/75

How similar are the perceptions of the different people? Since the information in a picture is only visual how can a person know about other sensual characteristics? Do the people disagree most on the visual aspects of the scene or the other sensory aspects? (If your friends are reticent to describe the scenes, or if they are unclear as to what you are getting at, ask them specific questions. For example, 'Is this a noisy place?' or 'Is this a windy place?')

3   Place yourself in a familiar environment. Your house, your shopping area or a nearby library would be good environments for this exercise. Close your eyes. Stay quietly in this one place for about five minutes. Pay attention to your surroundings. Then begin to describe your non-visual environment. If you have a tape recorder use it, or describe the area to a friend who can write it down, or type your description if you can type well enough without seeing.

Did you notice non-visual aspects of this environment that you did not notice when you had your eyes open? How would your description differ if you could see? If you did notice sounds/smells/etc. that you had not with your eyes open, does this mean that you had not used this information in your evaluation of the area while your eyes were open? For example, if the sounds had changed would you have noticed these changes? And would these changes affect your perceptions of the area?

4   'See' two movies or television shows. However, pay most of your attention to the sounds in the 'picture', especially to the non-dialogue sounds, the music in the background, the sound effects, etc. How important is this auditory information?

If it was done differently would the 'show' improve? Would it be worse? What specific auditory information was used well? Which was used poorly? How do the sounds contribute to the general impressions of the show?

Watch a television show without the sound on. How does this affect your perceptions of the show?

Now listen to a good radio show, a mystery or some story telling sequence. There is no visual information explicitly provided in this presentation because of the nature of the radio medium. How is visual information provided though? What techniques are used to implicitly convey visual information? Which techniques work well? Which work poorly? Explain using specific examples.

5   Develop an auditory score to accompany a visual display, or produce an object which accompanies a certain set of auditory information (i.e. an essay based on a piece of music, a form based on a sound, etc.).

2 Is there time in a picture?
or
How is process included in visual representations?

Figure 4

A painting is not only a final product. It is a product which was created over time. It was begun and it was finished. It existed in the painter's studio in many forms. It was a blank canvas, it was a few outlines, and it was a detailed and complete form. It changed its form while the artist painted slowly and thoughtfully, and when he would hurriedly sketch.

For many this entire process is invisible because they are unable to image dynamically within this medium. They may not have experienced painting a picture, nor viewed another person painting. They may therefore not have the appropriate *content* available in memory for interpretation. Alternatively, they may have never attended specifically to the *processes* involved in the creation, and therefore do not have adequate instructions available in memory for the production of the appropriate dynamic sequence. These individuals see only the final result of the process in a painting. However, people sensitive to the painting media can view time in a painting. They are able to comment on the process implicit in the painting, and they can point to areas where the timing is off and where the timing is handled well. They can recreate the production of the painting in their assessment of this object.

The Cubist painters explicitly represented time in a painting, by showing how a single object would be viewed at different times. This is a different sort of time in a visual experience. It was not time in the sequence of the artist's activities, but rather within the activities of the viewed object. Duchamp's *Nude Descending a Staircase* is an excellent example of this portrayal of time, as he shows figures in various positions in an overlapping pattern.

One need not confine oneself to paintings to view time in a picture however. It is possible to experience time in a picture of familiar area. For instance, in a picture of a set of buildings known well to an observer, he can pretend to walk around in the scene. He can place himself in the picture (in his mind) and move around the areas as though he were there. He can imagine going around corners and seeing things that are not visible in the picture plane. He can see what is behind him, and what is to the sides. He can recall being in the area, and can recreate familiar walks, or specific walks that are particularly vivid in memory. Though the picture of the buildings is static, the observer can add time dimension to the scene if he is appropriately experienced.* This is shown in the following example, a protocol of a person who was attempting to identify a picture of a building on her college campus:

> Oh, that doesn't look like a building I know very well either . . . it has a good chance of being the Commons building where the cafeteria is . . . except the cafeteria has little things that you can go out and sit on, but you usually get locked out when you go there, and that's probably what those things are, but I don't remember seeing them from below, don't really know what they looked like . . . now, behind, it is . . . a building that has little holes . . . and doesn't have any windows where the little holes are . . . um, now this building, no not this building, but 2C has little holes as you walk up the stairways, just like those little holes, uh . . . the question is whether that edge of the building has windows . . . . Well, let's think about where those funny balconies are on the Commons building, uh, they're not on the front where you walk in, you'd have to be facing the parking lot . . . uh, there's no parking lot back there as I look at the building. . . .

The *process* inherent in a visual representation is revealed in all three of these examples. They are available in the re-creation of a painting or the representation of different points in a sequence, or in the exploration of an environment. In the first and last cases, these changes over time are inaccessible to the inexperienced. Many cannot understand the timing in a painting because they are inexperienced in this area. Others cannot predict accurately what is around the corner in a picture, nor can they feel as though they are walking through the pictured area, because they are unfamiliar with the area. Similarly, the timing in a Cubist painting can only be appreciated by a viewer familiar with the portrayed object or sequence.

*Imagining walking around in an area is often a very effective retrieval cue for memories. If an observer forgets what a certain area looks like, he can often retrieve this information by imagining himself standing (or walking) in an area looking at his surroundings.

PROJECTS

1   Find a picture of an area you know well. A postcard of an area near your home, or a vacation picture from your own collection of pictures, will work well. Imagine what it is like to walk around in the pictured area.

Is this easy to do? Describe what you see as you walk around? What sorts of things are easy to see? Which are troublesome? Can you imagine what the picture taker would see if he turned around and faced the other direction? (If you practice this exercise you will become more proficient at it.)

Try these same things with a picture of an area with which you are *not* familiar. How do these two kinds of experiences differ?

2   Consider something which you have created. For example consider a painting, a dress, a house, a piece of sculpture, or a meal. Reconstruct the process involved in this creation. Imagine yourself painting the picture, making the movie, etc.

How well can you remember the process? How much of this is from 'general memory' and how much do you remember directly from looking at the art work? How much of your memory is of the sort 'I must have done X because I had just completed Y', and how much is specific, 'As I did X I remember the teapot boiling, and then as I began Y my child interrupted me'?

Consider something which you did not create, that you did not see while it was being formed. Try to imagine the process of creation inherent in this object. Is this easy or difficult? Why? Does this vary depending on whether or not you are familiar with the craft involved in the creation of this object?

3   Make an abstract line drawing. Begin in one place on a page, and move your pencil along the surface continuously.

Now, recreate the journey of your pencil in your mind. Pretend that you are walking along the line (you are miniaturized) and that you are looking around at the other lines. Describe what you see.

3 Do we view things all at once or do we view the world in pieces?

or

Do we view the world passively, or do we actively construct our concept of reality?

Figure 5

The area around me consists of a garden, a brick courtyard and a green building. However, in order for me to see this area I must look at the different elements separately. First I look at the edge of the building, then at the window, then at the courtyard, etc. Yet though I look at each element with a separate glance (a head movement or an eye movement) I see the scene as an integrated whole. Though I do not *see* all the elements together, in my mind I perceive my environment as a whole not as a set of unrelated details.

I can see the area as a unified area because I can see the location of each object and its relation to the other objects in view. If I were unable to see the relations between objects I would be unable to integrate the scene. For example, if I made a drawing into a jigsaw puzzle it would take a long time for me to put the entire scene together. If the pieces I cut were not mutually exclusive, if there was some overlap between the pieces, my task would be easier. I could use this rational information to integrate the separate scenes.

The visual system is constructed in such a way that it is constantly integrating small pieces of scenes. Only a small percentage of the information available at the retinal level is represented in a detailed way at the level of the cortex.* Only the small central area on the retina, the fovea, contains a dense set of receptors which provides for acuity in viewing. Non-foveal areas in the periphery of the retina convey only general information about general shape and movement of objects in the visual field. In addition, only a small portion of our surroundings is seen in one eye movement, and the eyes are continuously moving and sampling from the environment. There is a saccadic eye movement (a discrete jump from one fixation point to another) about four or five times a second,

*See Hochberg (1964) or Anderson (1975) or Lindsay and Norman (1972) or another basic physiological text for a detailed explanation of these levels.

and there are smaller movements more frequently. In fact, if a visual image is stabilized on the retina (with mirrors that move as the eye does) it will disappear (Pritchard, 1972).

Our visual system, at a basic physiological level, relies upon the integration of a large set of quick glances in order to identify and explore the environment.

At a cognitive level, Hochberg (1968) has demonstrated that human subjects can accurately integrate successively presented details (edges, corners, etc.) in order to judge the shape of an object. Subjects are particularly successful in this task if the information is presented rapidly and if relational information is effectively provided. An analogy to this process is seen at a larger scale in investigations of cognitive mapping (as will be discussed in Unit 14). We will most likely never see our entire neighbourhoods in one glance, unless we do a lot of flying or unless we see aerial maps. Yet we can conceptualize the whole neighbourhood and readily draw a map of it. If we have time to wander about in an area, if we have the chance to relate the various parts of this area, then we can form an impression of the integrated whole of that area. In order to relate this area to another, we must see how these two subareas are related, and the process continues as we come to understand a larger and larger area. One learns each area, and its relations to other known areas.

Let us see how this relates to the information we have stored in our minds. Make an image — a picture in your head — of the handle of the door on the left side of a specific car (the driver's side in America).

Where is the place for the key to fit in?

If you drew a line between this keyhole and the right rear tail light would this line go through the centre of the back seat?

If you drew a line between the left rear tyre and the front right headlight would this line go through the centre of the front seat (or between the seats if you have bucket seats)?

Think about how you performed these tasks. (Also check the car soon to determine the accuracy of your judgements.) How did the images vary to answer each question? How did you move from one image to another?

Introspective evidence indicate that images change drastically for each of these questions. To answer the question about the keyhole one must zero in on the image of the door handle. In order to answer the questions about the whole car the image must be enlarged, it must be blown up from a detail of the handle to a picture of the whole car. And in each image one must integrate small parts of the scenes, just as one must integrate parts of scenes when one is actually looking at them. This integration requires the expenditure of time and energy. For though you have information stored in your mind about the whole car, only a small bit of this is readily available to you while you are paying attention to specific details.

PROJECTS

1   Cut a small hole (about one inch in diameter) in a large piece of cardboard or thick paper. This will be your 'peephole'. Place a picture which you have never looked at beneath this peephole. Move the peephole along the surface of the picture, exploring its details. How many glances does it take to recognize the main elements of the picture? Is it easier if you move the peephole rapidly? Is it best to move the peephole continuously or in discrete steps (i.e. close your eyes, move the peephole, look, close your eyes, move the peephole, etc., as opposed to having a friend move the peephole while you are not looking)? How do you explain this?

Is it easier to move the peephole over the picture, or to move the picture behind the peephole? Why should these two techniques be any different?

Try these procedures for about seven or eight different pictures. Are some more difficult to identify than others. How do you explain this?

2   Walk around your house in your mind. Look out of each of the windows, and describe what you see in each view. Now, turn the house around on the lot. Make the side facing north face west, and the side facing south face east. Again, look out of each window, and describe the view which is now present.*

3   Look at a painting for twenty seconds and jot down your impressions of this painting? What do you see? What predominates in the painting? What do you like?

Now look at the painting for another minute, and again record your impressions. Do this again after another five minutes, and then an additional ten minutes.

How do your perceptions change? What is involved in this process of viewing? Can you discern any systematic changes.

*My thanks to Coeleen Kiebert-Jones for this exercise. My only hope is that she is able to forget it.

4 Do I see things as they really are?
or
Can I change the way I see things?

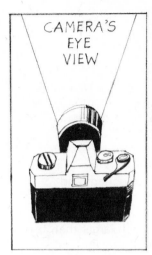

Figure 6

The question of how what a person sees in relation to how things 'really are' is a question that is deeply rooted in philosophic inquiry, one which provides the main context for perceptual psychological research, and one which provides the challenges within representational art.

When we realize that we do not see things as they really are, that we cannot generally tell how they really are, then we can constructively move beyond this view of the world and explore it anew. For the elusiveness of reality provides for the intrigue of illusions and for the concept of learning to see.

There are innumerable examples which demonstrate the lack of direct correspondence between objects in the world (distal stimuli) and the perceptions of these objects. (See Hochberg, 1964, for detailed summaries of these relationships.) The major examples that are investigated regard constancies and invariances. For instance, we experience constancies in relation to the size of objects. People do not get smaller as they walk away. Though the projection of objects on the retina does decrease in size as the objects move into the distance, we do not notice this change in size. What we perceive is not a direct mapping or the passive projection of objects on the visual system, but instead an active interpretation of all information available, including information from memory.

Another important constancy is shape constancy. Objects change shape continually as they are viewed from different angles or as they move relative to a stationary viewing position. Yet these changes are not generally noticed, and in general this information is discarded. One's house is the same house no matter from where it is viewed. A person playing volleyball is the same person as he moves around, though he forms different silhouettes against the sky. Similarly, a friend is the same person whether he smiles or frowns, even though his surface appearance changes. There are invariants in objects that we learn to attend to through experience, as certain variations in the perceptions of objects are judged of little importance.

A lack of direct correspondence between objects in the world and what we perceive also results from selective viewing. People are selective in viewing objects, as a function of what is judged important in a specific context. For example, one might ignore the smog in a mountain landscape, or the garbage cans in a picturesque alley. Similarly, one might concentrate on a redwood in a forest or a smile on a face. Selective viewing then affects the way new items are viewed, as distorted memory representations affect the encoding of new information, and the development of future expectations. Important and pronounced aspects of things can be exaggerated, a perceptual set that will often provide the basis of a good caricature but which will also cause inaccurate memories, as a person's ears may be seen as much larger than they are, or hair can be interpreted as being much longer than it truly is.

Figure 7

Figure 8

The world is not necessarily perceived as it really is. This allows one the flexibility of seeing in different ways, and in training oneself to be sensitive to certain aspects of the environment. When we look at the world we generally see objects. However we can change this if we concentrate. Try, for example, to see the space behind and between objects rather than the objects themselves. This negative space defines the shape of the objects directly, but it is experientially a different way to view the world. It stresses the 'ground' rather than the 'figure' in a scene (to use the terminology of the Gestalt perceptual psychologists). Now try consciously to change the way you see things by attending to certain colours or shapes in your environment. Take a quick glance around yourself, noticing only the things that are red. Take another glance and look only for blue things. Try this again, in the same area or other areas in your neighbourhood, as you move about during the day. With each glance change the sorts of things to which you attend. Look only for rounded objects; for pointed objects; for mechanical objects, for living objects; for man-made ugliness; for natural beauty. Each one of these viewing strategies will change your perception of the world around you. Even though certain information is always available in your environment new information

22

will appear to you if you view the world with different purposes. Your general perceptions and impressions will change. (See Leff, Gordon and Ferguson, 1974, for other approaches to this issue.)

PROJECTS

1   Photograph an object from different views. Get at least four different views. Now create a representation of this object which contains the information contained in all four pictures. For example, create another photograph or drawing which contains all four viewpoints, related to one another in specified ways.

2   Walk around in your present environment, for about five minutes, and touch objects in this environment. Then sit and contemplate those objects which you touched.

Describe the objects you touched. Why did you choose those and not the others? What attracted you to them? What kept you away from the other objects in the environment? Is colour the main determinant? Texture? Size?

Now, walk around the environment again, touching the same objects. Notice changes between your initial perceptions and these second impressions. Describe these differences.

Example

My present environment is an office which is a loft overlooking my living room. The loft has a redwood patio on the outside overlooking an ocean canal.

I touched almost everything; my teak desk top, wooden railing, carpet, ceiling, light, mobile, Japanese doll and teapot, candle, framed photographs, Monet poster under glass → outside; redwood patio, stucco exterior, wood railing, light, Japanese black pins.

The objects I touched were not selected, rather touch was almost exclusive. In some cases touch evoked a feeling of experience, newness, thought. Not really a profound revelation rather the realization that several objects here have existed without any personal involvement at all (touch, sight, smell, wonder, use...).

I guess the attraction, then, was thought provocation from involvement (newness) with an old, recognized object. Given this, all attributes played a role. None of dominant importance—colour, texture, size, shape.

Differences in perceptions were not specific, rather general thoughts. Thoughts about the role of objects in one's life, my involvement with them, foreignness of the immediate, lack of thought and interpretation of even the most basic and obvious of one's surroundings. (BW 10–75)

3   Find a picture which contains objects which vary in their distance from the camera. Trace these objects onto another piece of paper. Look at the sizes of the images in direct comparison, without the context of the images in direct comparison, without the context of the whole picture. Try this with a number of pictures, and attempt to find examples which are visually pleasing to you, and examples which are especially illusory.

4   Have some friends describe a certain place to you while they are in that place. For example, take some friends on a picnic and have them describe the area you are sitting in. Or find some friends in the store, take them to the same place in the store and ask them to describe it. Or have family members describe one of the rooms of your house.

You can ask them to write down their descriptions, or you can tape them. If you tape them, do this with one friend at a time, they should not hear each other's descriptions as this might effect their perceptions.

In addition to instructing your friends to describe the place, ask them to tell you: (1) which four things in their view are most important; (2) which four things are the most beautiful; and (3) which two things are the most angular. Systematically analyse your friends' descriptions, noting the similarities and the differences. (If you find it difficult to compare your friends' descriptions, ask them specific questions. For example give them a list of adjectives and ask them to check those which apply to the area they are viewing.)

5   Take some friends on a fifteen-minute walk around in your town. Secretly give each of these friends very different instructions about what they should be looking for on this walk. For example you might ask one friend to pretend he is an architect assessing the town for an architectural prize. Another friend might pretend he is thinking about opening a small business in the town and that he is deciding whether or not to go ahead with his plans. Another friend could pretend he is looking for a place to live.

Following this walk ask your friends a set of questions about what they saw. Use the same questions for all three friends. (It is best if none of the three are very familiar with the area you walk through then their answers will be based on what they saw on this one walk.)

How do your friends responses differ? Are there differences which are due to the differences between your friends? Are there differences due to the different sorts of instructions each friend was following? Give examples.

6   The worlds of objects, as we see it, is a three-dimensional world. We put together information about the locations of points in space to construct a view of objects. However, one can also 'see' the world in two-dimensional slices. You see the environment according to what you are looking for.

Look at the area around you, or move to another area which you find more interesting. Imagine a plane which is perpendicular to you, and five feet distant from you. Draw a line which shows objects which are on this plane. (This line traces the pattern your hand would follow if you touched objects on this plane. A picket fence would look like ᴧᴧᴧᴧᴧᴧ.) Try this for various slices of the area, planes which are at different angles and different distances.

5 Can you see what's in front of you and what's behind you at the same time?

or

Is our awareness composed of only that which we can see?

Figure 9

Take two minutes to look in your room, noticing the objects, their colour, their shape, and their size.

Now look straight ahead of you. Do not move your head. Is the room still the same?

Of course the room is the same you reply. Nothing has changed, except that you are looking straight ahead instead of all around yourself.

However, it is asked once again, since you cannot see behind you or to the sides, is the room for you at this moment not different than it was when you were looking all around?

Still you might insist that the room is exactly the same. Only what is seen has been changed. You are right of course. But re-examine your 'only' in the last statement. Do you really mean 'only'? Is what you see not a large part of your awareness?

Consider two statements: (1) This room has three tables in it. It contains a dining-room table, a desk and a coffee table. (2) When I look straight ahead I see only the desk.

When I look straight ahead the dining-room table and the coffee table still exist, though I do not see them. I could switch my glance and see them. They might exist in some reality, for example a friend might be looking at them right now. However, for me right now, they do not exist in my direct view. They do not exist 'as much' as the desk which I can see in front of me.

The coffee table and the dining-room table do exist in my mind however. I remember them. I know that they are behind me even though I cannot see them. I know this well enough so that I will be very surprised if I turn around quickly and find that they are not there. Similarly if I turn around and see a

25

pink sofa I will be very surprised. It is my memory of the area which allows me to feel the joy (or sadness) of surprise.

This does not mean that I have a perfect description of the dining-room table and coffee table in my memory, however. My descriptions may vary greatly from the real things. Yet they are probably adequate to allow the discrimination of these objects from many others which are similar. If you ask me to describe the coffee table I will not describe it as accurately as I would if I could look at it directly. I may forget elements of the graininess of the wood or the shape of the table legs, Yet this table does exist in my memory, and can be brought into my consciousness without my turning around.

PROJECTS

1   Imagine a world where things disappeared after you stopped looking at them. Imagine a world where you could not count on an object staying around if you turned your back on it. Create something which illustrates life in such a world (for example, a cartoon, a model or a short story).

A world like this is not so different than the world of our mental images. For example, imagine your mother's face. This image will stay in your consciousness only as long as you concentrate on it. When you stop concentrating it will disappear, and if you want to see it again you will have to reconstruct it.

Example
> What you don't realize is that such a world really exists. It's the world of immediacy. Memories must be sharper, more acute, because nothing exists after the first glance. You people on earth even attempt to describe this world inaccurately. In our world glances aren't 'objects' as you say, merely illusions. Objects are permanent, but here, as you say, nothing can be counted on to stay around.
>
> Memories to us are basic; another sense. We treasure them and the accomplishment of perfecting the storage of pictures. Sometimes it's far more peaceful, then, just closing our eyes. This is the only continuity in our world. You people on earth have things easy, you don't need or use the levels of memory you have, and we've reached. (BW 9–75)

2   Imagine a table top. Place a red book on this table, in your mind. Place a striped pillow on top of the red book. Place a record on top of this. Place a painting on top of this. Place a grapefruit on top of this.

OK? Are all items still in your image, or have some disappeared? If they have not disappeared, what have you done to maintain them? Try this same imaging exercise with other sets of objects. Are there any regularities in your ability to maintain images? For example, can you consistently maintain five things and no more?

3   Imagine a world where you had no memory of what you saw. In this world you would never know what to expect in the next glance. Take a moment to reflect on this situation, and then write a short description depicting a person who is experiencing such a world.

6 Is a photograph exactly like what we see?
or
Are we conscious of the visual relationships of the objects which exist on our retinas, or do we interpret these at such an early stage of processing that this information is, for all intents and purposes, unavailable?

Figure 10

We tend to assume that a photograph captures a view of reality. We often say that a painting looks so realistic that it may be a photograph and we set the photograph as a goal in our realistic drawing. If only we could make the pictured object look as real as a photograph we would be happy.

If this were a valid approach, then the art of photography would not exist. Photography would be simply a direct reflection of reality, and one would not need to be trained extensively to capture realism in a photograph. There would be no need to train people to previsualize, to predict how a picture will turn out, nor any need to train the photographer to take artistic photographs.

Consider Figure 11. This shows a picture of a building as 'we really see it'. As 'we really see' this building the top and bottom of the building converge, as a function of the angle of viewing and the distance relationships. Yet this kind of view does not capture our impression of the building. Even if the sides are non-parallel on our retinal images, we 'know' that they are parallel, and therefore perceive them as such. The photographic image is unable to reflect this complex knowledge, and hence does not make the necessary 'mapping' between the world as we see it and the world as we know it.

Another incongruity between a simple photograph and our view of the world can be discerned if you carefully consider all of the elements in your field of view. When we look at our immediate surroundings we also see our own

Figure 11

body. We see our chests, our legs and our arms.* These personal elements compose and frame a good part of our view, yet we rarely notice this. We make the necessary conclusions in our perceptions such that we can see our worlds independent of ourselves.

A further difference between photographs and our view of the world is a function of the boundaries of a photograph. Each segment of our view of the world consists of a small focused image and a large peripheral blurred area (in the foveal and non-foveal regions of the retina). We put a large set of overlapping focused images together to attain a cognitive perception of a scene. In contrast a photograph is a distinctly framed entity. It is rectangular, all of its regions are generally in focus, and there is no way to look beyond the bounds of the framing.

Another important difference between photographs and our view of the world involves the dimensionality of the images. Our view of the world is three dimensional. It deals with objects and locations which exist in three dimensions, and it is attained through binocular viewing, and motion parallax (resulting from movement of the head). Such is not the case with photographs. Photographs are two-dimensional entities. They are monocular representations of space and they are static. There surely are cues for three-dimensionality on the picture plane. Shading, occlusions and angularity allow us to interpret the picture as a three-dimensional area. However the three-dimensionality is not a property of the photograph, but instead of our perception and interpretation of the photograph.

PROJECTS

1   Take a photograph of a friend who is standing in front of you, and who had his hand stretched toward you. Take this photograph from very close to the person, being sure that all areas of his body are in focus. The hand which you photograph will seem much too large as compared with the size of your friend's face.

*Many artists play with this concept. Escher has created a particularly good representation of this.

Take three other pictures which demonstrate this same foregrounding effect. Explore this effect in a search for finding something which is pleasing to you, and in an attempt to make something appear humorous.

2   The general shape of a painting on the wall does not appear to vary when you look at it from different angles. Similarly, the shapes of objects in the pictured scenes do not appear to change with different viewing angles (unless the angles are very extreme). Walk around a painting, viewing it from a set of different locations. Now, view the painting from the same places (about five different views including extreme angles), and take pictures from these locations. Describe the distortions in these pictures, and explain why they are not present in normal viewing.

3   Choose some large complex pictures. Using four pieces of paper (or cardboard) as a frame, divide each large picture into smaller pictures in different ways.

What makes a certain way of framing better than another? How does the interpretation of certain parts of the pictures change with differences in framing?

7 How does what's in front of you compare with what's behind you?

or

Does memory directly represent the world as we see it?

Figure 12

A person's memory is the record of external events which she actively encodes in order to deal with future situations. It consists of certain content as well as sets of rules for dealing with this content.

Information in long-term memory (the general memory system) is that information which has successfully passed through earlier stages of processing, through basic perceptual feature analysis, through a sensory information store, through short-term memory and into long-term memory.*

Objects and events are initially represented by the memory in a sensory information store, a system which preserves most of the basic physical characteristics of an object which has disappeared. This sort of memory system is what is responsible for a person being able to remember a phone number by simply 'reading it from her mind'. For a short period after one is told a phone number one feels as though she can still hear it being said.

However, this information will quickly disappear unless it is rehearsed.

If information is rehearsed it will be maintained within the short-term memory system. The attributes of items stored by this system are more abstract than those stored in the sensory information store, yet they are not as abstract as those stored in the long-term memory. Short-term memory also differs from long-term memory in that it has a limited capacity of about seven items (see Miller, 1956), and in that errors in memory at this level will be acoustic rather than semantic.

*This explanation is somewhat simplistic, in that it presents only one description of the memory process and does not examine any alternatives. Also the many complications to this proposed structure are not discussed. See Massaro (1975) or Norman (1969) or Lindsay and Norman (1972) for a detailed description of memory.

In addition, information stored in the short-term memory system is very prone to being replaced by interfering stimuli unless it is encoded into long-term memory, or unless it is rehearsed.

In order for information encoded in long-term memory to be effectively used, it must be stored in a manner such that it can be retrieved accurately. Information which is not retrievable is non-existent for all intents and purposes. Numerous schemes of organizational strategies will allow for accurate retrieval. Hierarchical storage is effective, as is encoding based on linguistic labelling.

Images which relate objects, and spatial imagery also enhance organization and retrieval. Generalized network structures, which specify relations between concepts, are also effective.

It is important to note, however, that long-term memory is not necessarily a direct mapping of initial experiences. Organizational schemes at each level of memory affect the form of the data that is eventually retrieved. Information may be remembered inaccurately because of effective encoding at each level in the system, rather than simply because of failures in the memory system.

An object may be encoded as being round and small, for example, and later it may be impossible to recall whether the initial object was a tomato or a plum. Placing objects in classes, and relating them to other objects, may enhance recognition at a later time, but it may also result in inaccuracies in memory. These inaccuracies demonstrate the sorts of encoding mechanisms which have been found adaptive by a certain individual, and hence point to the workings of the memory system. However, such inaccuracies can also result in inappropriate stereotyping and prejudice.

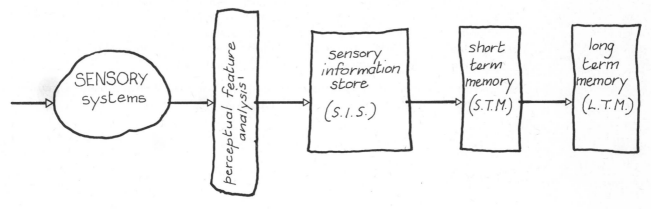

PROJECTS

1   Look at what is in front of you. Now turn around and describe what you saw.

Now turn back to the original scene. Look again. Turn around and again describe what you saw.

Continue this sequence. Each time you look at the area you will see more detail and add this to your description. Try this exercise with a number of different scenes.

Modify the exercise by drawing the scene rather than describing it. Again, you should see more detail in the scene each time you look at it, and you will begin really to see the objects you are looking at. (No excuses for your drawing abilities. You can notice how your drawing changes irrelevant to its basic appearance.)

Try one of these exercises again (describing or drawing). However, this time wait three minutes before you begin describing (or drawing) after you turn away from the scene. How does this affect your depiction of the scene? Does the waiting make the task more difficult? What do you find yourself doing during the three-minute delay? Dreaming idly? Madly going over the scene again and again in your mind? (If you are not doing the latter, try it. The rehearsal of the scene helps keep it 'alive'.)

Now consider these exercises carefully. How do your impressions of an area differ when you are looking directly at it and when you are re-creating it from memory?

Is your memory of an area just like a photograph? Explain.

Can you see details in this 'photograph'? Explain.

Do you look at this 'photograph in your head' in the same way you look at the real objects? If not, how does the process of viewing differ?

How can you describe your memory for scenes? How does your memory differ from the 'real thing'?

Systematically analyse the changes in your drawings and descriptions. How do they vary with repeated attempts, and with new glances at the area?

2   Modify the initial exercises slightly. Rather than beginning anew after each glance at an area, begin a description (or drawing), look back at the scene, turn around and continue the description (drawing), look again at the scene, etc. How many glances do you find yourself taking before you feel you have completely represented the area? Is this procedure easier or more difficult than the initial exercises? Use specific examples in your explanation.

3   Think about the upper corner of the room you are sitting in, without looking at it. What does it look like from the floor level?

Surely you would find it odd if this angle appeared in its incorrect form in your room, yet it may have been difficult for you to remember just how it appeared. Why do you think this is? Try this exercise with other angles. For example, imagine the angles on a certain window, or on an open door, or on a spiral staircase. Which angles are especially difficult to recall?

Are there many different classes of angles in your visual environment? Draw and describe instances of different types of angles.

8 Are there different styles of thinking?
or
Is visual thinking a legitimate and efficient kind of thinking?

Figure 13

In school we are taught how to think systematically. We are taught to solve problems and to complete designated tasks. We are taught to add, subtract, multiply and divide. We are taught to read and write. We are presented with millions of schemes for solving problems, from the scientific method to sudden insights. However, what is generally forgotten in all these presentations is an honest discussion of how we learn. We are taught (or we teach) in one way, but we may well learn in another. Making notes in margins of books, drawing explanatory diagrams, outlining underlining and imaging, are processes that play important roles in helping us to learn, yet these activities are rarely discussed in a classroom.

Why is this so? Why has there been so much more emphasis on teaching than learning?

We need to remember that education is a part of and takes place within the complex maze of society where extreme specialization and fragmentation have been the rule. As a result, schools have placed more and more emphasis on verbal concepts and linear thinking, while visual and holistic activities have been neglected.

As a consequence, instead of making progress in the development of fully functioning flexible thinkers, important capacities have rigidified or have been left undeveloped.

Hoping to develop optimal thinkers through such a one-sided approach to education is like trying to train an athlete to run the world's fastest mile while progressively cutting off pieces of one of his legs.

Robin B. Cano (1975)

33

The use of visual representations in learning, be it arithmetic, composition writing or anatomy, may be critical in allowing an individual to grasp a difficult concept.

> Practical drawings are mental tools. Once you have learned to make them, you will find that they are as useful in solving problems as saws and hammers are useful in carpentry.*

Visual representations provide a framework for visual thinking, an oftentimes powerful technique in attacking and clarifying a problem. Simple sketches and diagrams, and straightforward mental imagery, will often tame very wicked problems. Blackboards and sketchboards will often provide the context for important discoveries.

Techniques of visual thinking are generally set aside in favour of more formal procedures when we are taught in school. Of course formal techniques are often important, but it is imperative to remember that concrete examples and guides to visualization will aid in the comprehension of formal techniques. Surely thinking can be done in many everyday and academic situations without high levels of formal and symbolic manipulations. Moreover, non-formal reasoning may prove superior, as symbolic manipulations may not allow the necessary latitude in solutions to certain kinds of problems. This has been attested to be by numerous creative thinkers in our culture, including Albert Einstein, who explain their discoveries, thoughts, and inventions in terms of visual images. Images reported from dreams are oftentimes claimed responsible for creative thinking, as are shapes in the natural environment. Wholistic images are reported to provide transitions from sequential organizations which have been unsuccessful. (See Arnheim, 1969, and McKim, 1972, for an expansion of these basic ideas.)

In addition to the visual system, the kinaesthetic system often allows for a creative analysis of problems as it provides a mechanism for direct representation of spatial attributes of objects. People explain the form of certain objects by moving their hands in the air and they explain changes in positions of viewing by feeling themselves moving about. They also report recalling movement of an object (a person, a machine, etc.) in order to retrieve further information about that object. For example a person may try to imagine how an object falls in order to determine its shape, or how a person moves in order to recall his size. In fact, as spatial representation is a key to visual thinking, as well as pictorial representation, it may be that the kinaesthetic system is responsible in a large part for what is generally referred to as non-linear visual thinking.

PROJECTS

1   An important aspect of visual thinking is the ability to deal with spatial manipulations mentally, and to use these in solving problems. Figure 14. illustrates this.

Which of the five cubes can be made by folding the first pattern? Which of the four houses can be made by folding the initial pattern? (More than one may be correct in each instance.)

* From *Thinking with a Pencil* by Henning Nelmes, as reviewed on p. 357 of *The Last Whole Earth Catalog*, Penguin, 1972 (the set book for *Art and Environment*).

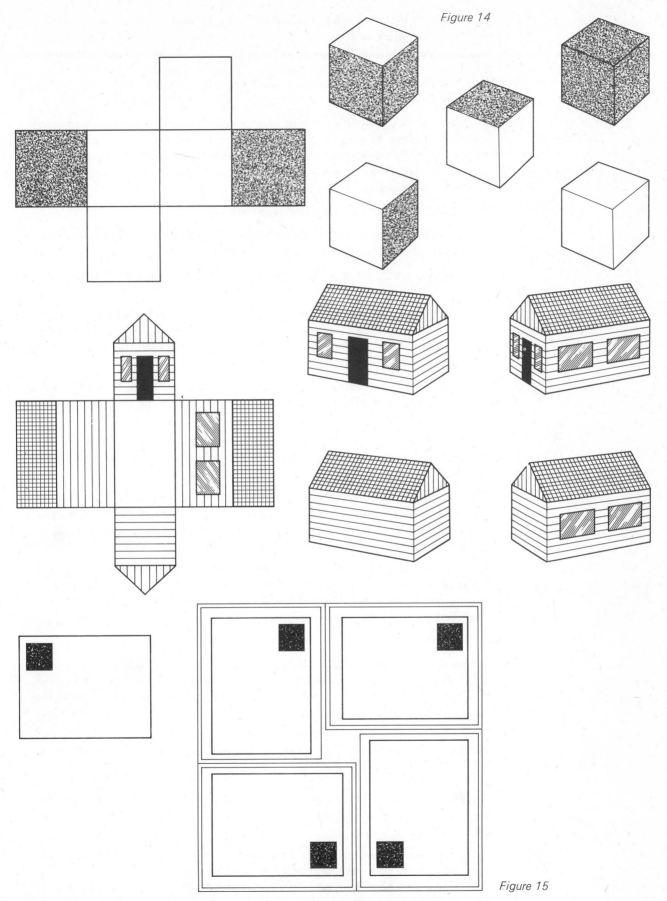

Figure 14

Figure 15

Consider Figure 15. Which of the four patterns are rotations of the initial pattern? Which require the pattern to be flipped into three-dimensional space?

How did you solve the problems in Figures 14 and 15? Which was the easiest? Why? Which was the hardest? Why?

Examples like these have been designed by psychologists to assess people's abilities to handle spatial relations. Do you think people vary much in their abilities to solve these problems? What professions might encourage the use of these abilities? Of what use are spatial relations to a person? Are there everyday tasks that require this sort of ability? How do you think these abilities relate to general intelligence?

2   There are basically two ways to give people directions about how to reach a certain location. You can explain the directions verbally or you can draw a map. In a similar fashion, when you receive directions you can either try to remember them verbatim or else make a mental map. Which of these descriptions describe your preferences?

As a direction giver? As a direction receiver?

Some people do not like to be given maps as directions, but would rather be given explicit verbal directions. Why do you think this is?

Other people prefer maps, and have difficulty remembering lists of verbal directions. Why do you think this is?

People can be classified into two groups as a function of their styles of thought. Visualizers are people who like maps, and verbalizers are those who like word lists. One can then give a person directions (teach them) differently, as a function of the individual's thought style.

However, we must not simply consider thought styles to be different for different people, but also must consider thought styles to be different for different situations.

For example if there are a lot of irregular hills in an area, a map often fails to represent all the critical areas. If there are few identifiable objects in an area it is sometimes difficult to represent the directions verbally.

This has profound implications for how we can effectively teach, how we can effectively learn, and how we can effectively think. We must have a large set of thinking tools available to use in a myriad of different situations. One of these techniques must be visual thinking.

Think of two examples of problems you have dealt with where visual thinking could be effectively used. Would most people have used this technique for solving these problems? Where did you learn to approach problems in this way? Did you ever learn this explicitly?

3   Use graphic notations to solve the following problem:

A man, a fox, a goose, and some corn are together on one side of the river with a boat. The goal is to transfer all of these entities to the other side of the river by means of the boat, which will carry the man and one other entity. The fox and the goose cannot be left alone together, nor can the goose and the corn.

Explain how your notational scheme helped or hindered your solution.

Would mental imagery have sufficed in the solution of this problem?

4   Imagine a 3 in × 3 in × 3 in cube which is painted red. Now slice this cube so that you end up with a set of one-inch cubes (this will require two slices on each side).

(a)   How many one-inch cubes are there?
(b)   How many have one, and only one, red side?
(c)   How many have two, and only two, red sides?
(d)   How many have three, and only three, red sides?
(e)   How many have no red sides?

Describe the thought processes you used to solve this task. Particularly note any kinaesthetic, visual or formal linguistic strategies that you used.

Figure 16

Process used in solving cube task Robin B. Cano

1   Trying to conceptualize cube—first as a whole with red paint covering the 'skin'. Then making the slices, first downward (= 3 sections) then horizontally — no, downward again in the other direction (now 9 sections) now cut horizontally so each of the 9s is cut into
3 = 9 × 3 = 27 one inch cubes.

2   Count each layer mentally to check this 1 layer = 3 + 3 + 3 = 9 cubes per layer. Each layer = same number so 3 × 9 = 27 OK.

(a)   *So number of cubes = 27*

3   (b)   *Only 2 red sides?* Corners would have 3 so not those. The only ones with 2 red sides would be those between the corners at the edge of the cube. On each side then there would be 4 such cubes. No, not on each side, only 2 sides because each is a 'wrap-around'. So think again in 3 slices. The two outside slices would each have 4 such cubes plus 4 more on the centre slice so $T = 8$. Yes, I think that's right — each 'in between' cube would have 2 red sides but be sure not to count twice going round corners.

(c)   *Only 3 red sides?* Corners. Yes, top and 2 sides. How many corners are there on a cube? 1, 2, 3, 4 on top layer, none in middle so 2 × 4 = *8.*

(d)   Oh, I left out how many have 1 red side. 1 in the centre of each face = 6.

(e)   No red sides = 1. Only central cube would have no connection with outside as the depth = 3 and 2 of these would have an outside surface. Yes 1.

4   Generalizing: I worked in small sections and then tried to figure out number of multiples.

5   I drew a model afterwards and did some counting trying to verify. Found counting more limiting than generalizing in some instances.

#8 Addendum to solution cube problem.

R. Cano

After working out solutions to the questions during a morning, I felt the problems "solved" and finalized. However, that was only my controlled thought at work. Approximately 10 hrs. later when I started to go to sleep for the night the thought of the "2-sides red" came into my mind and I knew I had made a mistake and where the mistake was located.

In a relaxed way I was able to visualize a view of the cube that allowed me to see 2 sides & the top. I realized that I had only counted the 2 end slices (1 cube deep) and that those 2 slices did indeed include 8 cubes with 2 sides red but that the central cubes on the center slice also had 2 red sides & that this added 4 more or a total of 12 cubes with 2 red sides.

It also appeared reasonable that there should be more cubes with 2 red sides than 3 red-sided cubes.

Thus resolved, I went on to more personal visualizing and hence to sleep. I felt more secure in this visually obtained data than in the logical generalization, which in turn was better than simple counting because it alone seemed to involve both sorts of checks — i.e. logical & visual. I found it reassuring & exciting to think about the capacity potential as related to learning, both for myself & for my students.

10/20/75

9 How can I best plan the rearrangement of the furniture in my room?
or
How can representational devices help in the production of good decisions?

Figure 17

The process of design, in architecture, engineering, interior decorating, or gardening, demands a certain level of trial and error. However, the number of trials (and errors) can greatly effect the resultant design, and can also effect the probabilities of completing a design. The issue for a theory of design is to determine a way to avoid a large number of cycles of trials and errors in the design process without sacrificing the flexibility important to good design.

In order to investigate this issue, consider how one might rearrange furniture in a room.

One technique is to imagine how things will look if they were moved. 'If I moved the desk against the wall with the window, and the couch near the bookshelves, then . . . but of course I could always do. . . .' The extravagant number of possibilities can cause even the best visualizer's mind to boggle at the imagery that is necessary in this simple task. One might be able to see the desk as it would look in its new location, and she might imagine how the couch would look under the bookshelves. However, these moves mean that other objects in the room (the chairs, the coffee table, etc.) would require moving. Yet, as it is difficult to hold all these parts of a room in your image of the room simultaneously, it is generally quite impossible to really get a very good general sense of how the entire room will appear after all the movings. Errors are therefore made in moving, many of which take great amounts of time and energy. (One can improve at this imaging task with practice. How good are you at this kind of imaging? Why do you think it is easy (or difficult) for you? Have you had much practice at similar tasks?)

Drawing helps one avoid the limited capacity of the imaging system. One can draw pictures of a room with the objects in different places, and one can draw a number of pictures in order to show all sides of the room. Similarly, one can cut out shapes which represent the size and shapes of the objects and move them around on a map of the room. In these ways one need not hold all the information about the room in her head, and can rather put this memory capacity to other portions of the task, such as evaluation.

Often individuals will purposively attempt to solve visual problems without using graphic notations, simply to show that they can visualize adequately. (This is similar to a child adding without using her fingers.) The absurdity of this is that great artists and self-proclaimed visualizers continually use explicit graphical representations in their work. They rely on their imaginations and visual images when this is most efficient. But they use external notations when they are useful. For example, Francoise Gilot described the process Picasso went through in painting her portrait:

> He painted a sheet of paper sky-blue and began to cut out oval shapes corresponding in varying degrees to this concept of my head; first, two that were perfectly round; then, three or four more based on his idea of doing it in width. . . . Then he pinned them onto the canvas, one after another, moving each one a little to the left or right, up or down, as it suited him. None seemed really appropriate until he reached the last one. Having tried all the others in various spots, he knew where he wanted it, and when he applied it to the canvas, the form seemed exactly right in just the spot he put it on. It was completely convincing. He stuck it to the damp canvas, stood aside, and said, 'Now, it's your portrait.'*

Of course the representation Picasso decided upon was his final product. In using drawings to represent a room you are not looking for the best drawing, but instead for the best room. A good drawing of a room propped up into three dimensions may be inappropriate as a real room in which people live.

Another technique available for the room problem is modelling. You can build scale models of the objects in a room, and move these about in a model of a room, systematically evaluating each arrangement. (This may not be sensible in deciding on the organization of a room. It may be far simpler to move the furniture. However, this kind of technique would be sensible if your task was the arrangement of larger items — buildings, trees, mountains, etc. — where the cost of an inaccurate movement is very high. Architects, for example, use models extensively in their work.) Of course even scale models present problems. You must still have to rely heavily on imaging. You must imagine yourself reduced to the scale of the model, and you must imagine yourself walking around in the space.

There is another approach which can solve the problem. If one is terribly ambitious one can make (or buy) a periscope (or modelscope) which allows her to see the model of the room at eye level from different places.†

However, when you reach this point you will probably decide to find a friend to help move the furniture around. For the direct movement of the furniture allows you to determine directly how the room will look and feel when it is arranged in different ways. You need not rely upon your imagination or your logic. No symbolic manipulations are necessary. You need only look at the room and see how you like it.

*Francoise Gilot and Carlton Lake (1964) *Life with Picasso*.

†This is the basic approach of the Environmental Simulation Project at the University of California at Berkeley. Architectural models are viewed through an optical probe, and presented to observers on film or videotape for evaluation. Plate 2(a) shows this system. The probe in this figure is $\frac{3}{8}$ inch in diameter. Plate 2(b) is an example of output from this system. Plate 2(c) is a photograph of the actual area.

Plate 2 (a)

Plate 2 (b)

Plate 2 (c)

Of course, what if four friends come to visit? How will the room appear when they are in it? How will it look to each of them?

Ah, is there no escape from using one's imagination?

An initial answer to this playful question is yes. Surely technologies can be developed which systematically generate and evaluate all possible design solutions. And in fact, many computer assisted design programmes have taken this approach, as have various rationalistic approaches to the theory of design. However, the problem with these approaches is that in order to specify truly an efficient design programme one must model the human mind — and this is no small task.

One must determine what the attributes are to which people attend in the evaluation process, and they must include these in a system which generates design alternatives. One does not want to consider all possible solutions to a design problem. Additionally, one wants to be able to generate solutions which are not confined to certain limited sets of parameters. One must allow the generation of creative alternatives, and redefinitions of problem spaces. Ah yes, there is probably no escape from using one's imagination. What a delightful situation!

PROJECTS

1   Design a small room, or set of rooms, for yourself. Initially use mental imagery and describe what the areas will be like. Then use drawings of different kinds (floor plans, perspective drawings, etc.). How do these tools help you?

Keep all the sketches and ideas you have. Analyse the progression of your ideas. Would a model help you in your designing?

In what ways?

Example
  Initially, I imagined a skeletal content-like design. What that told me was content, relative positions of objects and basic design of walls. After some more thought about space and dimensions I added design and dimensional characteristics (i.e. extended and expanded — even added — windows; added space by shaping the ceiling). What I did was add favourable attributes on top of basic content. I suppose a third phase, now, would be to re-evaluate the basic content in light of more desirable space and design attributes. Would a model help? What kind of model? One of thought, design and re-evaluation processes (sort of mental) *or* one where I could play around with various attributes, shapes, content items with an alterable structure 'mock house' model? Both! I suppose. (BW 9–75)

2   Design a room which could contain the following activities:

(a)   playing chess,
(b)   doing visual thinking exercises,
(c)   imaging,
(d)   talking with two or three friends,
(e)   writing,
(f)   drawing,
(g)   yoga.

Also provide at least one rounded wall surface, a way to fix snacks, and an entrance which one has to crawl through.

Now, analyse your design process. Explain how you reached your solution.

10 What kinds of information are used in drawing?
or
Why can't I draw?

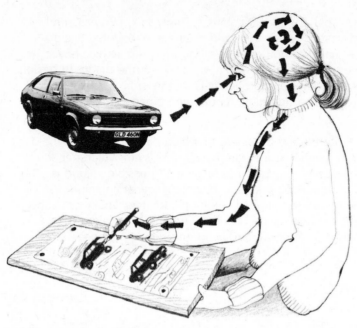

Figure 18

Some people seem to draw very well and very easily. Others are frustrated with their abilities and struggle desperately to produce even a simple sketch. Are some people not 'artistically inclined'? Are some people just unable to co-ordinate a drawing instrument? Is drawing a gift that some people have and others do not?

Each of these statements may have truth in them. However, a more constructive approach to the issue is necessary.

Drawing is a very complex activity (though simple in its very essence), that takes time, energy and a large number of mistakes in order to learn. It is an ever developing skill that changes and matures over time. One continually learns to use different media, and he becomes more and more sophisticated in seeing what is around him in different ways. A person has to learn to see carefully, he has to learn the conventions of the two-dimensional media to which we are all accustomed (whether or not we explicitly understood these conventions).

He must be able to represent perspective in scenes, a task which requires careful rendering of angles in the environment and superposition of objects. Additionally, size and shape constancies must be included in drawings, though variances from different viewing positions must also be included. Spatial relations within objects must be preserved, preventing distorted relations between these portions and other objects in the scenes.

Additionally, an important skill for the drawer, is to resolve the conflict between what is seen and what is known. It is difficult to draw a horse with only three legs, for example, because we know too well that horses have four legs. We also know that from some angles only three of the horse's legs will be visible. Yet this is easily forgotten in a rendering. Similarly, we can sometimes forget that from different positions the legs will appear to be unequal in length for we 'know' that they are the same length.

Some approaches to drawing attempt to integrate touching and seeing so as to help a person directly and carefully experience an object — its shape, texture and changes — so that he can draw it. Other approaches place value on a person becoming completely involved with his drawing. Others stress the importance of making errors in order to learn. All good approaches to drawing stress the interaction of numerous inputs, and the importance of slow and careful consideration of this activity as well as the rapid expression of simple attributes. Good drawing becomes more complicated than learning to see things as they are and learning to draw them as they are.

A person is effective in his drawing if he is able to know things about the world because of numerous sensory inputs, and he is able to represent these in the drawing media such that they can be interpreted successfully by another individual.

We interpret the three-dimensional world and we are generally unaware of two-dimensional images. Yet in drawing we must create a two-dimensional image which can be interpreted in three dimensions, a two-dimensional image which is not necessarily the same as that which we directly perceive at a retinal level. It is somewhere between this and the conceptual representation (knowledge) of the world we have stored in our memories.

Try the following exercise in order to explore the relationship between seeing and knowing. Take a transparent piece of material, put it in front of an object, and then trace the outline of this object. This exercise should convince you that in drawing we do not want an exact replica of an object, that we want a representation which conforms to our sense of what a drawing is. We want a representation of an object that captures its essence as we know it. We do not necessarily want a drawing of the world as we see it in a two-dimensional array.

The production of a drawing is a complex learned activity, as one must develop techniques which allow him to produce acceptable drawings as well as to develop simple ways to directly see the world as it is. However, the affirmative answers to questions at the beginning of this section are probably supported in an examination of people in our culture, as discouragement in one's drawing is a self-fulfilling prophesy. People who do not draw well do not ever draw, and hence they never learn to draw.

PROJECTS

1   Draw the front of your house from memory. Take about five minutes thinking about your house, and then fifteen minutes to sketch it.

Does this drawing look exactly like the front of your house?

Explain any differences and also describe similarities.

Were you limited by your drawing ability or by your memory? By both?

Explain.

Sit in front of your house, and draw it. How does this second drawing differ from the first drawing which you constructed from memory?

2   Look at a set of pictures quickly, then draw the basic shape of at least one of the pictures from memory. How does your drawing differ from the original pictures?

Describe the process of your drawing.

Look at the pictures again.
Draw (copy) the basic features of a few scenes while they are in front of you. How does drawing from a two-dimensional representation differ from a three-dimensional world?

How does this second picture compare to the first?

11   Is imagery alive and well?
or
Do some people have pictures in their heads and symphonies in their minds?

Figure 19

The Behavourists brought an end to the intellectual investigation of imagery in psychology in the 1920s. They stated quite clearly and simply that since an experimenter cannot see another person's images these images are not accessible to investigation. Unfortunately the psychology of this era was intimidated by these analyses, as the field was desperately seeking the status of a science, and did not want to be accused of any non-scientific endeavours. Imagery disappeared from academic psychology.

Recently an interest in mental imagery has been revitalized. Researchers have been actively discussing the role of imagery in thought and language, and research on memory has implicated the process of imagery in the effective storage and retrieval of information. (See Richardson, 1969, and Segal, 1971, for detailed discussions of current topics on imaging.)

It is still not completely clear to investigators what a person is doing when he forms a mental image, nor is it clear why an image should enhance memory. There are large numbers of hypotheses existing to explain images, yet it is unclear whether or not investigators will ever be able to tell which explanations are the most accurate, as images are very difficult to study definitively.

Fortunately for many, the experimental elusiveness of imagery need not interfere with experience of imaging nor with a fascination with these fleeting entities.

Experientially, visual imaging is a process which approximates viewing pictures in the mind. One need only concentrate and practice, and she can have a vivid display of images right before (behind?) her very eyes!

However, upon closer analysis imaging is found to be a complex process based on specific and general information, on concrete and abstract attributes, and on complex processes of organization. It is sort of like seeing a mental picture but it is different in very many ways. It is intertwined completely with memory, the content available there as well as the processes. One needs to know how to image as well as what to image, but if this information is available, imaging can provide for effective retrieval of visual and non-visual information.

Make an image (a picture in your mind) of a horse in a field. Take time to look carefully at this image and then answer the following questions based on an examination of your image:

(a)   Is the image in colour or in black and white?

(b)   Can you see all four legs of the horse, or is he standing such that one leg cannot be seen?

44

(c)   Is there a fence around the field?

(d)   Is there grass in the field?

(e)   Is it green or brown?

Are these questions easy to answer? Why or why not? Does it seem as though you are simply scanning the picture in your mind, as you would a picture in a book? If not, how does the process differ?

(There are no correct answers to these questions of course. In fact, the best answers to the questions are non-answers. For example, you might find that your image is not in colour, nor is it in black and white, or you might find that your horse has legs, but you cannot see them to count them, or that the grass in the field only has a colour after you read the question.)

Pretend you are standing in front of your house, at the main entrance. When you feel as though you are really there, answer the following questions.

(a)   Can you see the kitchen from this position?

(b)   Are there any people in your image?

(c)   How many windows can you see from this position?

(d)   How many doors?

(e)   How detailed is the image you generated for this task?

Someday soon go stand in the actual position in front of your house. Answer the same questions. How accurate were you when you were using the image of the area? How similar was the image you formed to the scene?

Now try these images.

Image a cow . . . . Image a building . . . . Image a Pinocchio . . . . Image a football player . . . . Image a whale . . . . Image a Rolls Royce . . . . Image a sunny day . . . . Image a cup of tea . . . .

How long does it take you, on the average, to make an image? Does it matter what you are trying to image, or do all images take the same amount of time? How do you decide when your image is complete?

Images need not be visual. Try these multi-sensory images. Take your time on them. Relax before you begin and rest between images. Do not simply do the images to do them, but rather let yourself become totally involved with each situation. If some of the images are particularly fun for you, take the time to express your image in another medium, for example, drawing or writing (see project 3 at the end of this section).

Imagine the smell of a friend's perfume.
Imagine the smell of a fresh peach.
Imagine the smell of freshly cut hay.
Imagine the smell of a barn.
Imagine the smell of a factory.

Imagine that you are in a swing, swinging high into the sky.
Imagine you are running as fast as you can.
Imagine you are taking a hot bath.
Imagine you are swimming in cold water.
Imagine the taste of a marshmallow.
Imagine the taste of lamb.
Imagine the taste of orange juice.

Imagine the sound of a car screeching to a halt.
Imagine the sound of laughter.
Imagine the sound of an organ playing in a church.

Imagine a fingernail scratching a blackboard.
Imagine a fast moving stream tumbling over boulders.
Imagine tomato juice pouring into a glass.

PROJECTS

1   Have some friends imagine the items in this last section. Then have a discussion about each person's images. Are images different for different people? Is it easy to talk about one's own images? Do your friends claim that they do much imaging in their daily lives?

2   If you were devoted to the experimental analysis of images, what might you do to convince the world that images exist and that they are worth systematic study?

3   Elaborate on three different images, which were discussed in this section, by expressing them in a medium of your choice.

4   Consider the multi-sensory images that you formed in this section. How did images in different sense modalities differ? Did images within modalities differ? Were some clearer and more vivid than others? Answer these questions using specific examples.

CONCLUSIONS

The following statements describe some of the issues raised in each section, in the discussion of the relation between our internal and external worlds. The number of each statement indicates the number of the section in which this topic was discussed.

1A   Visual information is extremely important in our assessment of our environments.

1B   The interaction of all of the sensory information in an environment is important. Though the visual aspects of an environment are generally primary, other sensory aspects of the environment contribute greatly to our perceptions.

1C   Non-visual information can be conveyed in a visual medium. Similarly, visual information can be inferred from non-visual media.

2A   Time is implicitly represented in a picture.

2B   Timing in a scene will be perceived differently depending on the experience of the observer.

3A   Seeing requires the integration of sets of discrete views.

3B   The explicit relationships between views aid in the cognitive integration of these views, which results in the phenomenological impression of complete scenes.

3C   Examining a mental image requires the movement of attention from one view to another.

4A   Our perceptions of the objects on the world are based on world knowledge, past experience, and our purposes in viewing, as well as the direct projection of images on our visual systems.

4B   You see the environment according to what you are looking for.

4C   Different people see things differently. The same person will see things differently at different times depending on his viewing strategy.

5A   Objects that we do not see exist in our primary consciousness. These objects are used in the interpretation of the world that we do see.

6A   Photographs reflect the world as we see it, not necessarily as we know it.

6B   Photographs represent direct mappings of the world, whereas we view the world in a very active, constructive, and interpretive fashion.

7A   What is behind you is not necessarily the same as what was in front of you.

7B   Memories for objects are systematically different from the original objects. Details are lost, details are added, and generalizations are constructed.

8A   Different thinking styles are appropriate for different levels of tasks and for different people.

8B   Visual thinking is oftentimes ignored as an appropriate technique for learning and solving problems, yet it is a technique which can be very effective.

9A   Imaging, drawing and modelling help a person predict the outcomes of the rearrangement of elements in a scene.

9B   The actual rearrangement of a scene, and the direct perception of the scene, provides the best basis for decisions about rearrangements. The effectiveness of representational media can be judged in terms of how closely they approach this realistic situation, and in terms of the ease of symbolic manipulations within the media.

10A Drawing is a complex activity. It is not simply a matter of learning to mirror the external world, but is instead an exercise in seeing the world and then communicating this in a way that other individuals will feel as though they are seeing the same world.

10B Drawing requires learning to see the essence of objects and learning to represent these in the drawing media using the implicit rules of the picture media.

11A Imaging is a process based on the information stored in memory and the creative combination of this information.

11B Imagery is not necessarily visual, and can be multi-sensory.

11C Images are elusive, and concretization of these constructs is difficult.

11D The impreciseness and vagueness of images, as the illusions in perception, provide us with a window on subjective reality.

I've moved through my mind many times in creating this journey for you... I've gone forward and backwards and sideways... I've looked at what I've learned about perception and thought from others, and I've explored things I have discovered myself. I've contemplated what I've known of creativity, and art, and joyous play. And, I've tried to look very carefully at my visual environment – my patio, my office, my living room, the Santa Cruz campus, the Sierra Nevadas, the Pacific Ocean, the mountains above the Russian River and the San Francisco skyline – in a way that I might communicate to you.

I've enjoyed my journey through my mind and through yours I've enjoyed taking the time to chat a bit in midair I hope that you were able to take the time to travel through your mind as well, and that you were able to enjoy it

KH - 1975

REFERENCES

Anderson, B. F. (1975) *Cognitive Psychology*, Academic Press.

Arnheim, R. (1969) *Visual Thinking*, University of California Press; Faber (1970).

Escher, M. C. (1967) *The Graphic Work of M. C. Escher*, Meredith Press; MacDonald (Ballantine, 1972).

Gombrich, E. H. (1960) *Art and Illusion*, Pantheon; Phaidon Press (1962).

Gregory, R. L. (1966) *Eye and Brain — Psychology of Seeing*, McGraw-Hill; Weidenfeld & Nicolson.

Gregory, R. L. (1970) *The Intelligent Eye*, McGraw-Hill (2nd edn, Weidenfeld & Nicolson, 1972).

Held, R. (ed.) (1974) *Image, Object and Illusion*: Readings from *Scientific American*, Freeman.

Hochberg, J. (1964) *Perception*, Prentice-Hall.

Hochberg, J. (1968) 'In the mind's eye', in Haber, R. N. (ed.) *Contemporary Theory and Research in Visual Perception*, Holt, Rinehart & Winston (new edn, 1970).

Hooper, K. (1973) Identification of Mirror Images of Real World Scenes, unpublished PhD dissertation.

Leff, H., Gordon, L., and Ferguson, J. (1974) 'Cognitive set and environmental awareness', *Environment and Behavior*, vol. 6, pp. 395–447.

Lenneberg, E. H. (1967) *Biological Foundations of Language*, Wiley.

Lindsay, P. H., and Norman, D. A. (1972) *Human Information Processing*, Academic Press.

McKim, R. H. (1972) *Experiences in Visual Thinking*, Brooks-Cole.

McLuhan, M. (1964) *Understanding Media*, Signet; Routledge & Kegan Paul (Sphere, 1973).

Massaro, D. (1975) *Experimental Psychology and Information Processing*, Rand McNally.

Miller, J. (1956) 'The magic number seven plus or minus two: Some limits on our capacity for processing information' *Psychology Review*, vol. 63, pp. 81–97

Nicolaides, K. (1971) *The Natural Way to Draw*, Houghton Mifflin; Deutsch (1972).

Norman, D. A. (1969) *Memory and Attention*, Wiley.

Ornstein, R. E. (1972) *The Psychology of Consciousness*, Freeman (paperback edn, 1973).

Pirenne, M. H. (1970) *Optics, Painting and Photography*, Cambridge University Press.

Posner, M. (1973) *Cognition: An Introduction*, Scott, Foresman.

Pritchard, R. (1972) 'Stabilized images on the retina', in Held, R., and Richard, W. (eds.) *Perception: Mechanisms and Models*: Readings from *Scientific American*, Freeman.

Richardson, A. (1969) *Mental Imagery*, Springer; Routledge & Kegan Paul.

Segal, S. J. (1971) *Imagery: Current Cognitive Approaches*, Academic Press.

Southworth, M. (1969) 'The Sonic Environment of Cities', *Environment and Behavior*, vol. 1, pp. 49–70.

Whorf, B. L. (1956) *Language, Thought and Reality*, MIT Press.

ACKNOWLEDGEMENTS

Frontispiece, Kent Marshall; Plate 1, Janie Rhyne; Plates 2a, 2b, and 2c, University of California, Berkeley, Environmental Simulation Project; Figures 1, 2 and 11, Kristina Hooper; Figures 16 and 19, Dana Cuff.

THE ART AND ENVIRONMENT COURSE TEAM

Simon Nicholson (*Chairperson*)
Christopher Cornford
Christopher Crickmay
Barry Cunliffe
Katherine Dunn
Kristina Hooper
Richard Orton
David Stea
Susan Triesman
Janet Woollacott

John Barr (*Course Assistant*)
Phillipe Duchastel (*IET*)
Gerald Hancock (*Staff Tutor*)
Pam Higgins (*Designer*)
Caryl Hunter-Brown (*Liaison Librarian*)
John Kenward (*Assistant Staff Tutor*)
Maureen MacKenzie (*Secretary*)

BBC

Richard Callanan
Andrew Crilly (*Co-ordinating Producer*)
John Selwyn Gilbert
Noella Smith
Nat Taylor
Nancy Thomas

Programme Consultants

Hugh Davies
Mark Francis
Edward Goldsmith
Roger Hart
Bernard Leach
Ray Lorenzo
Chrissy Maher
Marion Milner
Eric Mottram
Stephanie Roberts
Ken Worpole

'The Great Divide' Collective

Kathy Henderson
Maureen McCue
Michelene Wandor

ART AND ENVIRONMENT

Unit 1   The empty box
Unit 2   Our conversation with things and places
Unit 3   Natural sound
Unit 4   Art and everyday life
Unit 5   Imaging and visual thinking
Unit 6   The great divide: The sexual division of labour, or 'Is it art?'
Unit 7   Social relationships in art
Unit 8   The moving image
Unit 9   Verballistics
Unit 10   Interactive art and play
Unit 11   Electronic sound
Unit 12   Body, mind, stage and street
Unit 13   Boundary shifting
Unit 14   Environmental mapping
Unit 15   Design with nature?
Unit 16   Art and political action